ALGEBRAIC MANIPULATION OF OPERATORS

ALGEBRAIC MANIPULATION OF OPERATORS

THE FOUNDATION OF MATHEMATICS SUCCESSES

Jayeola S. Muyiwa

Jayeola S. Muyiwa

ALGEBRAIC MANIPULATION OF OPERATORS

THE FOUNDATION OF MATHEMATIC SUCCESSES

Published by Spines Publishing Platform

ISBN: 979-8-90222-147-0

Spines

ORDERING INFORMATION:

To order this book, please write to:

info@davedansolutions.com

YouTube: davedansolutions–Mathematics with JAYEE

Foreword

Algebra stands as one of the foundational pillars of mathematics, shaping the way we express patterns, relationships, and abstract ideas. Its language allows us to move beyond numbers into the realm of general reasoning, bringing abstract with the world of reality. This book has been written with clarity, precision, and a deep appreciation for both the beauty and utility of algebraic operators. It guides readers step–by–step from fundamental concepts to more advanced ideas, encouraging thoughtful engagement rather than rote memorization.

Whether you are encountering algebra for the first time or rediscovering it with fresh perspective, this text invites you to explore the subject as a living discipline—one that continues to evolve, inspire, and solve real problems. I strongly hope that following this book, with particular attention on working examples and exercises, will not only strengthen your mathematical skills but also deepen your curiosity and confidence.

Dr. Michael Olatunji

This book is dedicated to:

The Trinity (God the Father, Jesus Christ, the Savior and the Holy Spirit).

My Wife (Oluwakemi Jayeola) and Children.

Acknowledgments

I want to express my profuse and profound gratitude to Dr. Bassey Akpan for his timely support by providing useful information for the realization of this book. I also appreciate Dr. Philip Adebo for the invaluable support given.

Contents

INTRODUCTION: Algebraic Manipulation of Operators

Algebraic manipulation of operators is a foundational skill in high school mathematics that allows students to simplify, rearrange, and solve algebraic expressions and equations using the four basic arithmetic operations: addition (+), subtraction (–), multiplication (×), and division (÷). Mastery of these operations and their properties enable students to handle more complex mathematical problems involving functions, equations, inequalities, and calculus. Understanding algebraic manipulation not only strengthens problem–solving skills but also develops logical thinking, precision, and analytical reasoning—skills that are crucial in science, technology, engineering, and mathematics (STEM) disciplines (Kieran, 2007).

At its core, **algebraic manipulation** involves applying arithmetic operations to variables and constants according to specific algebraic rules. These operations are guided by fundamental laws such as the **commutative**, **associative**, and **distributive** properties. For instance, addition and multiplication are commutative, meaning that the order of the terms does not affect the result:

$a + b = b + a$ and $a \times b = b \times a$.

Similarly, they are associative, meaning the grouping of terms can change without affecting the outcome:

$(a + b) + c = a + (b + c)$.

The distributive property links multiplication and addition, allowing expressions to be expanded or factored:

$a(b + c) = ab + ac$.

These properties form the algebraic backbone for simplifying and solving expressions in high school mathematics (Van Amerom, 2002).

These operations are also foundational in working with inequalities, exponents, radicals, and logarithms. Moreover, algebraic manipulation underpins key topics like coordinating geometry (e.g., finding slope or intercept), calculus (differentiation and integration), and physics (formulas involving force, motion, and energy).

Furthermore, the ability to manipulate operators helps students model real-world situations mathematically—such as calculating profit, growth rate, or chemical reaction rates. By understanding operator manipulation, students can express relationships between variables and predict outcomes through equations and formulas (Sfard, 1991).

In summary, algebraic manipulation of operators—addition, subtraction, multiplication, and division—is a fundamental component of high school mathematics education. It equips students with the tools to handle abstract problems, express mathematical relationships clearly, and apply logical reasoning. Beyond academics, these skills contribute to everyday problem-solving and analytical decision-making. The effective teaching of algebraic manipulation prepares students for higher-level mathematics and scientific disciplines, reinforcing the broader goal of mathematics education: developing critical thinkers capable of tackling complex problems.

1
Addition And Subtraction Of Numbers

Addition and subtraction are two of the most fundamental operations in mathematics. They are the building blocks for arithmetic and algebraic reasoning, forming the basis for more advanced mathematical concepts taught in higher grades. **Addition** is the process of combining two or more numbers to find their total or sum while **Subtraction** is the process of finding the difference between two numbers. It shows how much one number exceeds another. In practical terms, **addition** is used when quantities are combined or increased—such as finding the total cost of items—while **subtraction** is used when quantities are reduced or compared—such as finding how much money remains after a purchase. Together, these operations form the foundation for everyday calculations and advanced problem–solving in mathematics. In arithmetic, 2 - 1 *is* 1, *but* 1 – 2 , is impossible. This is the smallest challenge among others such as question posts to students in high school and college. In this chapter, we shall offer a strategy that has made many students who have learnt under us successful over the years.

Worked Example 1.1

1 - 2 = ?

The number, 1 – 2 can be separated as 1 and –2. Each number carries either a positive or negative sign. Consider the number that has positive sign (+) before it as what you have (own) and the number that has *negative sign* before it as a debt (owe). Let us apply the terms own and owe to 1 - 2 = –1 . This implies you have 1 (own) and owe 2; then you still owe 1. Mathematically, 1 - 2 = –1 . The negative before 1 is a sign you owe.

Worked Example 1.2

–1 - 2 = ?

We would apply the same strategy to this question. Both numbers have negative sign before them, that is –1 and –2. If we apply the strategy used in the first example, it means we *owe 1* and *owe 2*; our debt is 3 (–3). The negative symbolizes it is a debt.

Worked Example 1.3

–6 + 10 = ?

Before we answer this question, we believe you should have understood the strategy. You owe (debt) 6 and own (have) 10. This implies you have 4.

Worked Example 1.4

12 - 20 = ?

The above question tells us that you own (have) 12 and owe (debt) 20. Then, you still owe 8. Therefore, 12 – 20 = –8.

Summary of the Strategy

- Separate the numbers into negative and positive
- Tag the number that has negative sign before it as what you owe and the number that has positive sign as what you own (have).
- Then, ask a question, "I own and I owe". The outcome depends on the numbers.

Exercise 1

Simplify the following

a. $-5 - 2 = ?$

b. $30 - 52 = ?$

c. $-12 - 20 = ?$

d. $-10 + 15 = ?$

2
Addition And Subtraction Of Alphanumeric

An alphanumeric expression consists of both letters and numbers combined through mathematical operations such as addition (+), subtraction (–), multiplication (×), or division (÷). The letters in these expressions are known as variables and are used to represent unknown quantities.In high school algebra, students are introduced to the manipulation of alphanumeric expressions — that is, expressions containing both numbers (numerical coefficients) and letters (variables). The addition and subtraction of such expressions are fundamental operations that form the basis for solving equations, simplifying expressions, and understanding more advanced algebraic concepts. These operations follow systematic rules that govern how like and unlike terms can be combined.

In this chapter, we would employ the strategy we use in chapter one. Remember, negative sign refers to what you owe (debt) and positive sign is what you own (have). The letters, which are also called variables, are the additions to the strategy in the preceding chapter. The major challenge to many students is how to handle the variables in the addition and subtraction of alphanumeric. *Here is what to do! Consider any variable (letter) as a currency such as dollars, naira, pounds etc. once they are similar in the expression.*

Worked Example 2.1

$2x - 4x = ?$

Let us separate them into $2x$ *and* $-4x$. Now, consider x as "dollar". Using the strategy of *owe and own*, it means you own (have) 2 dollars because it is the number that has a positive sign (+)before it, and You owe (debt) 4 dollars because it has a negative sign (–) before it. It is concluded you owe 2 dollars. You would replace dollar with x :algebraically, the result is $-2x$.

Worked Example 2.2

$-2x - 5x = ?$

By separating them, we have $-2x$ and $-5x$. It implies you owe (debt) 2 dollars and owe (debt) 5 dollars, considering x as dollar; the result is that you owe 7 dollars. It would be $-7x$.

Worked Example 2.3

$-10y + 3y = ?$

This time, we would apply the strategy directly to the question. Remember letter y would be considered as dollars. $-10y + 3y$ is termed to owe (debt) 10 dollars and own (have) 3 dollars. This means you owe 7 dollars. You would remove "dollars" and replace with "x" while writing the final answer, that is $-7x$.

Worked Example 2.4

$-5y^2 + 2y^2 = ?$

Recall that once the variables are the same in terms, apply the strategy. We have y^2 in both numbers, then y^2 our dollars. $-5y^2 + 2y^2$ is considered as you owe 5 dollars and own 2 dollars. The result is that you owe 3 dollars. Now, you want to write your final answer, remove dollars and replace with y^2.Algebraically, it is $-3y^2$.

Exercise 2

1. $2f + 5f =?$
2. $-a - 2a=?$
3. $-2y^2 - 4y^2=?$
4. $10v^2 - 6v^2=?$

3
Multiplication Of Numbers

Multiplication is one of the four fundamental operations in arithmetic and algebra. In high school mathematics, it serves as a foundational skill that extends beyond simple number operations to include algebraic expressions, polynomials, and functions. Understanding multiplication allows students to apply mathematical reasoning in a variety of contexts such as geometry, physics, economics, and computer science. Multiplication is a mathematical operation that represents repeated addition. Multiplication can be applied to whole numbers, decimals, fractions and integers. This chapter would only deal with whole numbers and integers. Multiplication is not only an arithmetic operation but also a key component in understanding higher–level mathematical concepts. In algebra, it forms the basis for working with **polynomials**, **functions**, and **quadratic equations**. In geometry, multiplication is used to calculate **areas, volumes, and scaling factors**. In statistics, it assists in calculating **probabilities** and **weighted averages**. According to Larson and Boswell (2017), mastery of multiplication enables students to transition from concrete arithmetic reasoning to abstract algebraic thinking, which is crucial for success in advanced mathematics and real–world problem–solving.

The rules of signs are essential when multiplying integers:

- Positive × Positive = Positive
- Negative × Negative = Positive
- Positive × Negative = Negative
- Negative × Positive = Negative

Common Mistake and How to Avoid It

Students sometimes make errors during multiplication due to:

Ignoring the sign rules: forgetting that a negative times a negative gives a positive.

Worked Example 3.1

$-2 \times -3 = ?$

We would take the following steps while solving a problem like this.

Step 1: Separate the numbers and the signs then multiply.

$(-) \times (-) = (+)$

$2 \times 3 = 6$

Step 2: Combine your answer in step 1 together.

$-2 \times -3 = +6$

Worked Example 3.2

−5 × 6 = ?

Step 1: Separate the numbers and the signs then multiply.

(−) × (+) = (−)

5 × 6 = 30

Step 2: Combine your answer in step 1 together.

−5 × 6 = −30

Worked Example 3.3

5 × −8 = ?

We would apply the strategy to the last two questions directly.

(+) × (−) = (−) **and** 5 × 8 = 40. Therefore, 5 × −8 = −40

Worked Example 3.4

$-4 \times -3 = ?$

$(-) \times (-) = (+)$ **and** $4 \times 3 = 12$. Therefore, $-4 \times -3 = +12$

Exercise 3

a. $-14 \times -3 = ?$

b. $-8 \times 3 = ?$

c. $-20 \times 4 = ?$

4
Multiplication Of Alphanumeric

Multiplication of alphanumeric expressions is a key concept in algebra that builds upon students' understanding of numerical multiplication. Unlike simple arithmetic, alphanumeric multiplication involves numbers (coefficients) and letters (variables), requiring the application of specific algebraic rules and properties. This operation forms the foundation for higher-level mathematical concepts such as expanding brackets, simplifying polynomials, and solving quadratic equations. An **alphanumeric expression** is an algebraic statement that combines numbers and letters using mathematical operations. The letters represent

variables (unknown quantities), and the numbers are **coefficients** that indicate how many times the variable is used. Multiplication of these expressions follows algebraic rules based on combining like terms, using exponents, and applying the distributive property (Hornby, 2019).

Worked Example 4.1

$-2y \times -3y = ?$

Step 1: We would apply the strategy used in chapter 3 by separating the numbers and the signs. In addition, we would separate the variable.

$(-) \times (-) = (+)$

$2 \times 3 = 6$

$y \times y = y^2$ (add the index/power/exponent of each together).

Step 2: Combine your answer in step 1 together.

$-2y \times -3y = +6y^2$

Worked Example 4.2

$-4g \times 5g = ?$

By applying the strategy in example 1,$-4g \times 5g$ becomes

$(-) \times (+) = (-)$

$(4 \times 5) = 20$

And $(g \times g) = g^2$

Therefore, $-4g \times 5g = -20g^2$

Worked Example 4.3

$6f^2 \times 4f = ?$

The two terms have positive sign before them. Positive times positive is positive; $6 \times 4 = 24$; $f^2 \times f = f^3$. Thus, $6f^2 \times 4f = 24f^3$.

Exercise 4

1. $-7f^2 \times 5f = ?$
2. $10a^2 \times -6a^2 = ?$
3. $4y^2 \times y^2 = ?$
4. $-8a^2 \times -6a^3 = ?$

5
Division Of Numbers

Division is one of the four fundamental operations in arithmetic, alongside addition, subtraction, and multiplication. It is the process of determining how many times one number (the divisor) is contained within another number (the dividend). The result of this operation is called the quotient, and sometimes there is a remainder if the division is not exact.

Division can be represented as:

$$\text{Dividend} \div \text{Divisor} = \text{Quotient}$$

or

$$\frac{\text{Dividend}}{\text{Divisor}} = \text{Quotient}$$

The rules for signs are:

- Positive ÷ Positive = Positive
- Negative ÷ Negative = Positive
- Positive ÷ Negative = Negative
- Negative ÷ Positive = Negative
- **Note:** Division **by zero is undefined.**

Worked Example 5.1

−20 ÷ −5 = ?

Step 1: Separate the numbers and the signs like we have done in the previous chapters.

(−) ÷ (−) = (+)

20 ÷ 5 = 4

Step 2: Combine your answers.

$-20 \div -5 = 4$

Worked Example 5.2

$-60 \div 20 = ?$

Apply the two steps directly.

$(-) \div (+) = -$ and $60 \div 20 = 3$.

Thus, $-60 \div 20 = -3$

Exercise 5

a. $-40 \div -5 = ?$

b. $-32 \div 8 = ?$

c. $20 \div -4 = ?$

6
Division Of Alphanumeric

Alphanumeric expressions (also called algebraic expressions) are combinations of numbers (numerical coefficients) and letters (variables) connected by arithmetic operators such as +, −, ×, ÷ . The division of alphanumeric expressions involves finding how many times one algebraic expression (the divisor) fits into another (the dividend). It extends the concept of numerical division by applying the laws of exponents and algebraic simplification. According to Charles et al. (2018), algebraic operations, including division, form the foundation for manipulating variables and constants to solve equations and model real–life situations. The **division of alphanumeric expressions** means dividing both the **numerical** and **literal (letter)** parts of a term or expression. It follows the same principle as numerical division but adds rules for handling variables.

If we have two algebraic terms:

$$\frac{a^m}{a^n} = a^{m-n}$$

and

$$\frac{ka^m}{la^n} = \frac{k}{l} \times a^{m-n}$$

where *k* and *l* are coefficients (numbers), and *a* is a variable.

This principle is based on the **law of indices (exponents)**, which states that when dividing powers with the same base, we **subtract the exponents** (Van de Walle, Karp, & Bay–Williams, 2019).

Worked Example 1

$$\frac{-20x}{-4x} = ?$$

Separate them into three, signs, numbers and variable. Then combine your answers.

Then we have

$\frac{(-)}{(-)} = +$, $\frac{20}{4} = 5$ and $\frac{x}{x} = 1$ (subtract power of the variable x)

Therefore, $\frac{-20x}{-4x} = +5.$

Worked Example 2

$$\frac{20y^5}{-10y^2} = ?$$

Apply the same strategy used in worked example.

$$\frac{(+)}{(-)} = (-),\ \frac{20}{10} = 2 \text{ and } \frac{y^5}{y^2} = y^{5-2} = y^3$$

$$Thus, \frac{20y^5}{-10y^2} = -2y^3$$

Worked Example 3

$$\frac{-12x^5y^3z^{-4}}{-4x^3y^2z^{-2}} = ?$$

We can break it into chunks and simplify.

$$\frac{(-)}{(-)} = (+),\ \frac{12}{4} = 3,\ \frac{x^5}{x^3} = x^{5-3} = x^2,\ \frac{y^3}{y^2} = y^{3-2} = y,\ \frac{z^{-4}}{z^{-2}} = z^{-4-(-2)} = z^{-2}$$

By combining each answer from the simplification, we have $\frac{-12x^5y^3z^{-4}}{-4x^3y^2z^{-2}} = 3x^2yz^{-2}$

Exercise 6

a. $\frac{-50x^5}{-10x^3} = ?$

b. $\frac{-16x^5y^3}{4x^{-3}y^{-2}} = ?$

c. $\frac{-35x^5y^3z^{-4}}{-7x^3y^2z^{-2}} = ?$

7
Solving Simple Equations

Solving simple equations is a foundational concept in mathematics generally. It involves finding the value(s) of an unknown variable that makes the equation true. Equations form the basis for more advanced mathematical concepts used in science, engineering, economics, and everyday problem–solving. An **equation** is a mathematical statement that asserts the equality of two expressions. It usually contains one or more variables.

For example:

$$y + 2 = 5$$

In this equation, y is the unknown variable, and the goal is to find the value of that makes both sides equal.

According to **Kirkwood & O'Sullivan (2021)**, an equation can be described as a balance between two expressions where mathematical operations are used to find unknown quantities while maintaining equality.

Basic Steps in Solving Simple Equations

To solve a simple equation, the following steps are commonly used:

1. **Identify the variable** – Determine the unknown you are solving for.
2. **Simplify both sides** – Combine like terms and simplify arithmetic operations on both sides.
3. **Isolate the variable** – Use inverse operations (addition ↔ subtraction, multiplication ↔ division) to move all

terms involving the variable to one side and constants to the other.

4. **Check the solution** – Substitute the value back into the original equation to verify it makes both sides.

Checking and Verification: Always check your final answer by substituting the found value of the variable back into the original equation. This ensures no mistakes were made in algebraic manipulation (Sullivan & Struve, 2020). Verification reinforces understanding and accuracy.

Worked example 7.1

$5x + 4 = 19$

To isolate x , Subtract 4 from both sides

$$\begin{array}{r|l} 5x + 4 & = 19 \\ -4 & -4 \\ \hline 5x & = 15 \end{array}$$

To solve for *x* divide both sides by 5.

$$\frac{5x}{5} = \frac{15}{5}, \; x = 3$$

Checking and Verification of answer

$5x + 4 = 19$

Plug in the value of *x* in the equation.

We have, $5(3) + 4 = 19$

Worked example 7.2

$-3y - 3 = -15$

To isolate *x*, add 3 to both sides

$$\begin{array}{r|l} -3y - 3 & = -15 \\ +3 & = \;\; +3 \\ \hline -3y & = -12 \end{array}$$

To solve for *x* divide both sides by −3.

$$\frac{-3y}{-3} = \frac{-12}{-3}, \ y = 4$$

Worked Example 7.3

−7*a* - 3 = 2*a* + 15

Collect the like terms by subtracting 2*a* and adding 3 to both sides.

$$\begin{array}{l|l} -7a - 3 & = 2a + 15 \\ -2a + 3 & -2a + 3 \\ \hline -9a & = \quad 18 \end{array}$$

To solve for *x* divide both sides by −9.

$$\frac{-9a}{-9} = \frac{18}{-9}, \ a = -2$$

Exercise 7

Solve the following Equations

a. $-2c + 11 = 3c - 4$

b. $5f - 16 + -5f + 4$

c. $12 - 6x = 7x - 14$

d. $8g + 14 = 6g + 4$

e. $23 - 6z = 16z - 43$

8
Solving Simple Exponential Equations

Solving simple exponential equations requires we find the value of the unknown variable wherever it appears in the equation. However, simple exponential equations can appear in different forms; it depends on the formation of the question. This chapter deals with two common forms.

a. Solving exponential equations in which the unknown appears as the base on one side of the equation. This type demands a lucid understanding of exponent rules and inverse operations. These equations often take the form

$$a^x = b^y,$$

where is the unknown base. To solve the equation, we need to isolate the base (making the base to stand on its own) by multiplying the powers of both sides by the reciprocal (inverse) power of the side with unknown base.

Worked Example 8.1

What is the value of y in $y^{\frac{1}{2}} = 2^{\frac{3}{2}}$?

Step 1: Identify the unknown variable. In this case, "y" is the unknown variable.

Step 2: Identify the power/exponent of the identified unknown variable, y; it is $\frac{1}{2}$.

Step 3: Obtain the reciprocal/inverse of the power $(\frac{1}{2})$; the inverse is $\frac{2}{1}$.

Step 4: Multiply the power of both sides with the reciprocal, $\frac{2}{1}$.

Step 5: simplify

Then we have, $\boldsymbol{y^{\frac{1}{2}\times\frac{2}{1}} = 2^{\frac{3}{2}\times\frac{2}{1}}}$

$y = 23$

$y = 8$

Worked Example 8.2

What is the value of $g\ in\ g^{\frac{3}{5}} = 3^{\frac{9}{5}}$?

We would now apply all the steps used in the worked example

1.

Step 1: Identify the unknown variable. In this case, "g" is the unknown variable.

Step 2: Identify the power/exponent of the identified unknown variable, y; it is $\frac{3}{5}$.

Step 3: Obtain the reciprocal/inverse of the power $(\frac{3}{5})$; the inverse is $\frac{5}{3}$.

Step 4: Multiply the power of both sides with the reciprocal, $\frac{5}{3}$.

Step 5: simplify

Then we have, $g^{\frac{3}{5}\times\frac{5}{3}} = 2^{\frac{9}{5}\times\frac{5}{3}}$

$$g = 3^3 = 3 \times 3 \times 3 = 27$$

Worked Example 8.3

What is the value of x in $x^{\frac{2}{3}} = 6^{\frac{-4}{3}}$?

We would follow all the steps used in the previous examples without repeating them. The equation becomes:

$$x^{\frac{2}{3}\times\frac{3}{2}} = 6^{\frac{-4}{3}\times\frac{3}{2}}$$

$$x = 6^{-2}$$

$$x = \frac{1}{6^2} = \frac{1}{36}$$

Worked Example 8.4

What is the value of f in $f^{\frac{-3}{7}} = 4^{\frac{-9}{14}}$?

We would follow all the steps used in the previous examples without repeating them. The equation becomes:

$$f^{\frac{-3}{7} \times \frac{7}{-3}} = 4^{\frac{-9}{14} \times \frac{7}{-3}}$$

$$f = 4^{\frac{3}{2}} = (\sqrt{4})^3$$

$$f = 2^3 = 8$$

Exercise 8a

What is the value of unknown variables in the following equation?

$y^{\frac{2}{5}} = 4^{\frac{6}{5}}$ $b.\ m^{\frac{-5}{12}} = 5^{\frac{5}{-6}}$ $c.\ z^{\frac{-2}{3}} = 25$ $d. (\sqrt{h})^5 = 16^{\frac{5}{4}}$

b. Solving unlike exponential equations where the unknown is in the exponent. These equations commonly take the form $x^a = y$, where and a are known numbers and is the unknown. The basic strategy is to rewrite both sides of the equation to the same base or to apply logarithms when rewriting is not possible. When both sides of an equation can be expressed as powers of the same base, the exponents can be equated directly otherwise use natural logarithm (base 10).

Worked Example 8.5

Solve for the unknown in $3^x = 9$

As stated in the passage above, express both sides of the equation to the same base. $9 = 3 \times 3$ and it can be written in power form as 3^2.

Therefore, $3^x = 3^2$.

The base cancels out, thus $x = 2$

Worked Example 8.6

Solve for the unknown in $5^{-y} = 25$.

We can rewrite 25 to its prime factor as $5 \times 5 = 25$. Remember, we need to make the base of both sides homogenous.

Thus, $5^{-y} = 5^2$

When the base of both sides cancels out, we have

$-y = 2$

$y = -2$

Worked Example 8.7

Solve for the unknown in $3^{-\frac{g}{3}} = \frac{1}{81}$.

Prime factor of 81 is $3 \times 3 \times 3 \times 3 = 3^4$.

We have, $3^{-\frac{g}{3}} = \frac{1}{3^4}$.

By law of reciprocal/inverse, $\frac{1}{3^4} = 3^{-4}$

Therefore, $3^{-\frac{g}{3}} = \frac{1}{3^4}$ is now $3^{-\frac{g}{3}} = 3^{-4}$

When the base cancels out, $-\frac{g}{3} = -4$

Multiply both sides by -3

$g = 12$

Worked Example 8.8

$6^{-p} = (36)^{-\frac{3}{2}}$. What is p?

$36 = 6 \times 6 = 6^2$

$6^{-p} = (6^2)^{-\frac{3}{2}}$.

$-p = (6)^{-3}$

$-p = \frac{1}{6^3}$, $-p = \frac{1}{216}$

Thus, $p = -\frac{1}{216}$

Exercise 8b

Solve for the unknown in the following equations

a. $2^{-y} = 4$

b. $4^{b} = 64$

c. $(\frac{5}{2})^{-y} = (\frac{25}{4})^{-1}$

d. $(\sqrt{3})^{x} = \frac{1}{9}$

Vocabularies

Alphanumeric: combination of letters (alphabets) and numbers (numerals).

Base: The number that is multiplied by itself when raised to a power (exponent).

Complex: Involving multiple steps, concepts, or strategies to find a solution, often requires critical thinking, logical reasoning,

Equation: A statement that shows two expressions are equal. It uses an equal sign (=) to indicate that the value on the left–hand side is the same as the value on the right–hand side.

Homogenous: Uniform or the same throughout.

Lucid: Clear, easy to understand, and logically expressed.

Manipulation: The process of rearranging, simplifying, or transforming mathematical expressions or equations to make them easier to understand or solve.

Operator: Tells you what action to take in a mathematical expression – such as adding, subtracting, multiplying, or dividing.

Power/Exponent/Index: All refer to the same idea: the number that tells how many times a base is multiplied by itself. These words are often used interchangeably.

Reciprocal: The number you get when you invert a value—that is, when you flip a number over.

Strategy: Planned method or approach used to solve a problem effectively.

Variable: A symbol (usually a letter) that represents an unknown or changeable quantity.

Answers To Exercises

Exercise 1

a. -7

b. -22

c. -32

d. 5

Exercise 2

a. $7f$

b. $-3a$

c. $-6y^2$

d. $4v^4$

Exercise 3

a. 42

b. -24

c. -80

Exercise 4

a. $-35f^3$

b. $-60a^4$

c. $4y^4$

d. $48a^5$

Exercise 5

a. 8

b. -4

c. -5

Exercise 6

a. $5x^2$

b. $-4x^8y^5$

c. $5x^2yz^{-2}$

Exercise 7

a. $c = 3$

b. $f = 2$

c. $x = 2$

d. $d = -5$

e. $z = 3$

Exercise 8a

a. $y = 64$

b. $m = 25$

c. $z = \frac{1}{125}$ or 5^{-3}

d. $h = 4$

Exercise 8b

a. $y = 2$

b. b = 3

c. $y = 2$

d. $x = -4$

Bibliography

Charles, R., Hall, B., Kennedy, D., Bellman, A., Bragg, S., Handlin, W., & Murphy, S. (2018). Mathematics Course 2. Pearson Education.

Hornby, A. S. (2019). Oxford Advanced Learner's Dictionary (10th ed.). Oxford University Press.

Hornsby, J., Lial, M. L., & Rockswold, G. (2017). Algebra for College Students (10th ed.). Pearson Education.

Kieran, C. (2007). Learning and teaching algebra at the middle school through college levels: Building meaning for symbols and their manipulation. In F. K. Lester (Ed.), Second Handbook of Research on Mathematics Teaching and Learning (pp. 707–762). Information Age Publishing.

Kirkpatrick, L., & Renshaw, S. (2021). Algebra for High School Students. McGraw–Hill Education.

Kirkwood, J., & O'Sullivan, A. (2021). Elementary Algebra. OpenStax, Rice University.

Larson, R., & Boswell, L. (2017). Algebra 1: Common Core Edition. Houghton Mifflin Harcourt.

Lial, M. L., Hornsby, J., & Schneider, D. (2019). Mathematics: Basic Concepts and Applications. Pearson Education.

National Council of Teachers of Mathematics (NCTM). (2020). Principles and Standards for School Mathematics.

Sfard, A. (1991). On the dual nature of mathematical conceptions: Reflections on processes and objects as different sides of the same coin.Educational Studies in Mathematics, 22(1), 1–36.

Stewart, J., Redlin, L., & Watson, S. (2016). Algebra and Trigonometry. Cengage Learning.

Sullivan, M., & Struve, K. (2020). Algebra and Trigonometry (11th ed.). Pearson Education.

Usiskin, Z. (2015). Reconceptualizing Algebra Education in the 21st Century. Information Age Publishing.

Van Amerom, B. A. (2002). Reinventing early algebra: Developmental research on the transition from arithmetic to algebra.Eindhoven University of Technology Press.

Van de Walle, J. A., Karp, K. S., & Bay-Williams, J. M. (2019). Elementary and Middle School Mathematics: Teaching Developmentally (10th ed.). Pearson.

About the Author

Jayeola S. Muyiwa is an accomplished educator and researcher in the field of Statistics and Data Science. He obtained a Higher National Diploma (HND) in Statistics from the Federal Polytechnic Ado-Ekiti, Nigeria. He also obtained a Bachelor of Science (BSc) degree in Statistics with first-class honors from Joseph Ayo Babalola University, Ilesa, Nigeria. To satisfy his academic thirst, he proceeded to the University of Malaya, Kuala Lumpur, Malaysia, where he earned a Master of Science (M.Sc.) degree in Statistics. Having arrived in the United States, he successfully earned a master's degree in data science with a PhD grade from the University of North Texas, Denton, USA.

His teaching career spans over 25 years, during which he has taught at middle, secondary and tertiary levels, inspiring students to develop logical reasoning and analytical skills through Statistics, Mathematics and Data Science. His motivation for writing this textbook stems from a burning desire to address the foundational challenges of students observed in the classroom over the periods he has taught Mathematics and Statistics in Africa and the USA. When not teaching or writing, he enjoys mentoring young educators and exploring the use of technology in education.

www.ingramcontent.com/pod-product-compliance
Lightning Source LLC
LaVergne TN
LVHW022314200926
848481LV00004B/1336